A YEAR ON THE DESERT

Second edition

For Jack, who loves stories,
and for his dad, Brian,
who loves Jack, and loves
reading stories to him;

Many happy memories
to you both!

[illegible]
March, 2020

A YEAR ON THE DESERT

Second edition

by Barbara Goodheart

illustrated by Mel Hunter

Best eWay Publications, Inc., Fort Lauderdale, Florida

To: Ken, Karen, and Diane

A Year on the Desert
by Barbara Goodheart, illustrated by Mel Hunter

Printed in the United States of America
ISBN: 978-0-9910062-3-6

PREFACE

To My Readers

Often it takes the efforts and passion of a group of people—not to mention a lot of luck—to bring out a book. That's especially true of this book, *A Year on the Desert*—first published decades ago by Prentice-Hall as a hardcover book. So I wanted to share with my readers the story of the book's original creation—and its resurrection.

How It Began

When I was a young writer, freelancing articles for magazines, I attended the Off-Campus Writers Workshop in Winnetka, Illinois. The group rotated topics and instructors, and I submitted a manuscript for each genre—from poetry to fiction.

Elisa Bialk, a noted author of children's books, taught a six-week session on writing for children. I submitted what is now Chapter II of *A Year on the Desert*—A Summer Night in Death Valley—as an article to send to a children's magazine, my first try at writing for children.

After reading my submission to the group, Ms. Bialk said, to my surprise, "This isn't an article for a children's magazine. It's a chapter in a children's book."

She suggested I send the draft to Prentice-Hall, calling it a book chapter, along with a query letter and a list of titles for additional chapters. I did, and soon received a contract from an editor there, Jean Reynolds.

How It Ended

A friend, Sally Olds, a fellow member of the American Society of Journalists and Authors (ASJA), read a prefinal draft of the book, encouraged me to go ahead, and offered helpful suggestions.

Soon after I submitted the first draft, Prentice-Hall accepted *A Year on the Desert,* without revisions.

A year later I had a go-ahead for a second book from Prentice-Hall, this one on the animal life in Florida's Everglades. But a difficult personal situation came up, making other commitments necessary, so, with deep regret, I had to bow out.

That didn't bode well for the desert book. As they tend to do today, readers were turning away from stand-alone books. And so, without a second book in the series, *A Year on the Desert* eventually went out of print.

. . . "It Reads Like Poetry" . . . "It's Not Just for Children" . . . "Adults Love It, Too". . .

Over the years I heard these comments from various readers, among them friends who were fellow members of ASJA—Richard Dunlop, Elliott McCleary, and Bonnie Remsberg. I respected their opinions, and regretted that the book was no longer in print.

I thought about the school children I'd visited, and how they'd enjoyed hearing the stories I'd read them from the book. I remembered a teacher whose students I'd visited;

she told me that a little girl in her class who wanted to be a writer was sleeping every night with my autograph under her pillow.

So, the desert book remained in my thoughts. But the problems and marketing challenges that go along with a stand-alone book seemed too difficult . . .

Until, one day, Amazon launched its internet-based technology.

And, suddenly, bringing a book out as an ebook and a print-on-demand paperback became not only possible, but feasible.

Hurdles

Two issues remained. Several fellow authors had suggested replacing the black and white drawings with color art or color photos—more modern, they said; wider appeal; it's what readers, especially young readers, look for. I almost took their advice, even though I loved the existing black and white illustrations.

I turned to my husband, Clyde, for advice. We explored the backstory of the artist, Mel Hunter, and became intrigued—and more reluctant than ever to part with Mel's drawings.

A Suggestion . . . and One More Hurdle

Our granddaughter, Nanci Goodheart Repala, and her husband, Greg, were helping us with reissuing the book, and they suggested keeping the existing artwork—"Not only because it's beautiful, but to make sure the illustrations stay in the background. Instead of drawing attention, they leave the emphasis on the story itself. Color illustrations would draw the reader's interest away from the story, and focus it on the artwork."

They were right, of course.

Now one last hurdle remained. The artist—actually, the artist's widow—held copyright to the drawings. We had hoped to reissue the book as a labor of love, not as a money-maker, and paying royalties to reuse the drawings of a famous artist could be prohibitively costly. And Mel had indeed become famous; his lithographs were now quite expensive.

We delved into Susan Smith-Hunter's background. She, too, was a talented artist. She was also a ceramic sculptor and art educator, with a Master of Fine Arts degree from Tulane University.

Clyde phoned Ms. Smith-Hunter. When he described *A Year on the Desert,* and our plans, she immediately and graciously granted us the rights to use Mel's artwork.

We are deeply grateful to her. *A Year on the Desert* would not be the same without Mel's illustrations.

In Closing

Many people, some no longer with us, made it possible to publish *A Year on the Desert* years ago, and to reissue it now for new generations of young readers.

I especially want to thank Elisa Bialk, Richard Dunlop, Elliott McCleary, Bonnie Remsberg, Jean Reynolds, Sally Olds, Nanci Goodheart Repala, Greg Repala, and Susan Smith-Hunter.

Also, the organizers and members of the Off-Campus Writers Workshop, in Winnetka, Illinois; the people at Prentice-Hall; and the staff at Amazon, especially those who developed the technologies that have enabled authors to share their works with readers.

I'm especially grateful to the late Mel Hunter for his appreciation of wildlife and his love of nature, so very evident in his inspiring artwork. And, again, to Susan Smith-Hunter, for so generously allowing us to reuse Mel's illustrations.

And above all I want to thank my husband, Clyde. His patience, understanding, and support seem to know no end. This book wouldn't have been possible without him.

—Barbara Goodheart

Fort Lauderdale, Florida

July 10, 2017

CONTENTS

	Preface	i
I.	The Story of the Spadefoot Toads	1
II.	A Summer Night in Death Valley	9
III.	The Bighorn Sheep	16
IV.	Cloudburst!	24
V.	Autumn	30
VI.	Winter	37
VII.	Spring	40
VIII.	The Springtime of the Birds	46
	Epilogue	53
	Bibliography	54
	The Author and the Artist	56

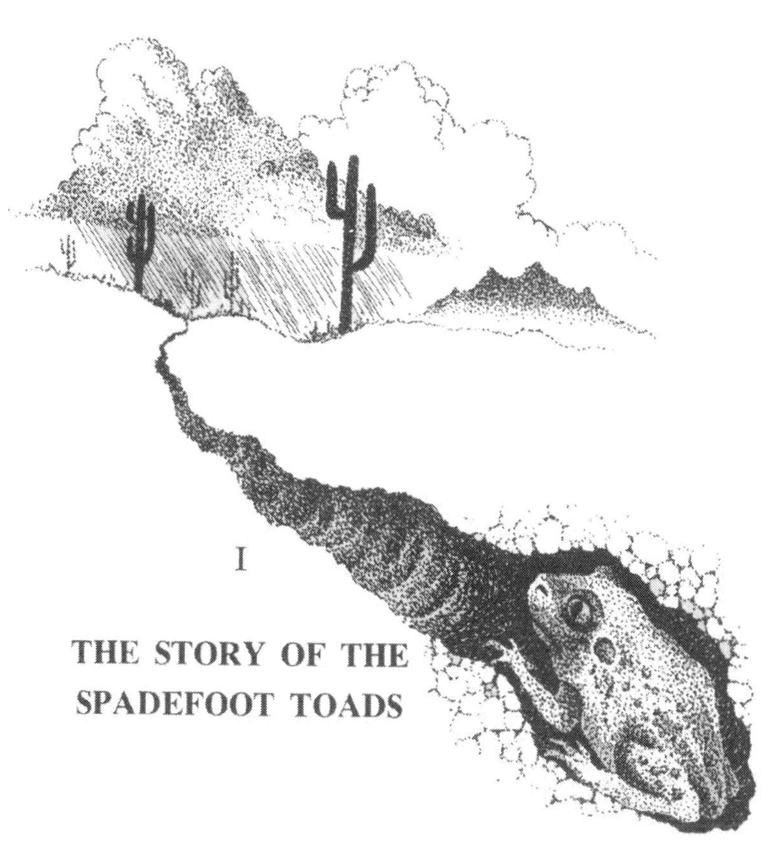

I

THE STORY OF THE SPADEFOOT TOADS

It was July in southern Arizona's Sonoran Desert. The wind picked up a bit of loose sand and tossed it in the air; a few tiny grains bounced off a saguaro and dropped silently to the ground. On a nearby slope an ocotillo plant swayed, its bare stalks reaching toward the sky like long, bony arms.

Suddenly the wind became stronger and the sky darkened. The temperature dropped below one hundred degrees for the first time since morning. A long streak of lightning cut across the sky toward the west, then abruptly changed direction. Thunder shattered the immense desert silence.

Then the rain hit, and it hit hard; huge, heavy drops thumping the sand and pressing small plants flat against the ground.

A trickle of water crept along a stretch of sand, then picked up speed as it rushed downhill. Nearby, deep in the ground, a Couch's spadefoot toad sensed the rain and stirred. He had huddled in his burrow for an entire year, ignoring the rain that had fallen in September and February. But now he squirmed restlessly. This was the summer rain, and the urge to be up and out was too strong to ignore. He must go!

Dusty and shriveled, his yellowish-green skin partly covered with dark, almost cocoon-like patches of dead skin, he made his way up toward the burrow entrance. Bracing his long-unused muscles and lowering his bony head, he butted the sand that blocked the entrance and broke through to the outside.

Like a hard shower, the rain pounded his skin. The toad sat a moment, motionless, as, sponge-like, he pulled the moisture in through his skin. This was his way of drinking. Some of the water would be stored in his lymph spaces and in his bladder; someday it might save his life, for the nearest permanent water was many miles away.

Two hours later the rain stopped. Three inches had fallen—a tremendous amount for one storm, for only seven inches had fallen on that part of the desert during the entire year.

Now it was night, and scattered pools dotted the sand. The toad could not see the large rain pool that had formed in a low area several dozens of yards away, but he sensed that it was there and hopped over to it. It was six inches deep and perhaps ten feet across. The toad entered the water and rested a few moments, his nose and his bulging eyes just above the water. Suddenly the toad shuddered and drew a tremendous amount of air into his vocal sacs until his throat was huge and swollen, like a balloon. He passed the air back and forth between his lungs and vocal sacs. As he did so, his vocal cords vibrated, and a melodic “baa” filled the air.

The toad repeated the sound every few seconds. It was clear and resonant, and it carried well. Dozens of yards away others of his kind heard his call and headed toward the rain pool.

A young female reached the pool ahead of the others and approached the singing toad, gently brushing against him. Turning, he climbed on her back and embraced her, his front legs clasping her body just in front of her hind legs. She uttered no sound. She was larger than the male—about three inches long—and her body was swollen with hundreds of eggs. Because of this, and because of her silence, he did not let her go. Nor would he, while she was plump with ripe eggs. Not even death would break his hold.

In response to his embrace, the female toad released her eggs. The male shed his sperm over them, fertilizing them. Soon others of their kind made their way to the rainpool and found mates. During most of that night the toads' lamb-like songs filled the air.

Shortly before dawn the toads left, one by one, to search for insects. A dozen or so yards from the pool a toad caught sight of a fly and watched, still as a rock, as it circled closer and closer.

The fly landed on the branch of a creosote bush and crawled toward the tip. Now! Flicking his sticky tongue lightning-fast, the toad caught the fly.

Just then a startled jackrabbit tore by, his long ears flattened and his slim, strong legs pushing into the sand. Instantly the toads hunched up and lowered their heads, as if by not seeing they could avoid being seen. Although the noise and movement had frightened them, they were not in danger from the jackrabbit. The real threat, at that moment, lay as still and silent as death. It was a young rattlesnake, motionless behind a bush.

The rattlesnake was hungry. Slowly he raised his head and flicked his forked tongue, testing the air for odors. Silently he began uncoiling.

A few yards away sat an old male toad, his farsighted eyes fixed on a string of ants off in the distance. He took three hops toward them. As he moved the ground vibrated ever so slightly. Although the rattlesnake could not hear airborne sounds, his ear bone picked up the ground vibrations.

Suddenly the young rattlesnake caught sight of the toad and slithered quickly after him, his rattles silent. Just as the old toad hopped again, the snake lunged. Faster than the eye could follow, the snake rotated his maxillary bones forward, bringing his fangs into an erect position at the front of his mouth. Instantly he sank his fangs into the toad.

The old toad gave a strange cry and struggled frantically, but the snake held on. Moments later the toad, paralyzed by venom, ceased struggling. The rattlesnake opened his mouth very wide and began swallowing the toad. Many minutes later he headed toward a plump cactus, where he would leisurely digest his meal. It was a good meal for the young rattlesnake. He would not need to eat again for perhaps a week.

It was becoming quite light, and the toads faced a new danger: the hot sun. Less able than mammals to adjust to high temperatures, the toads could quickly die from the heat. Or they could die from water loss—an animal's big danger in the hot desert, where little or no replacement water is available. The toads' skin offered them no protection against water loss. In an emergency, they could use their stored water, but then, unless they found replacement water, they would die.

They would be safe deep in the ground, where the air is much cooler and damper. And so—just a day after they had burst through the sand—the toads began to dig. Using the large, horny, shovel-like projections on their hind feet—the "spades" for which they were named —the toads tossed pebbles and sand in all directions. Rapidly scooting backward, they disappeared, covering themselves with sand as they went. In a few moments they were gone, each in his own burrow deep in the ground.

However, the toads had not left the desert quite as they had found it, for not far away, in the rain pool, little masses of jelly-covered eggs clung to small plants just under the surface of the water. Chances were slim that the eggs could develop into toads, for even if the eggs hatched, the tadpoles would need water for a long time or they would dry out and die. And a rain pool does not last long on the Sonoran Desert in mid-summer! That very afternoon the temperature rose to one hundred and five degrees and the water level dropped noticeably.

Yet the little toad eggs were developing rapidly. Each egg had already divided, first into two cells, then four, then eight, sixteen, and finally into many cells. On the second day, a curious thing happened: the eggs twitched a tiny bit; then, later, they began to jerk every which way like dozens of jumping beans stuck together, trying to move in different directions. The tiny embryo toads developing inside the egg capsules were trying to get out. They needed more oxygen. Soon, too, they would need food, for the yolk that had been nourishing them was almost gone.

That afternoon a miraculous thing happened: a tiny tadpole wriggled through his egg capsule and jelly coat and swam free in the water. Another followed, then another. In less than two days the eggs had developed into tadpoles.

The tadpoles rested for a day or two, then began to swim about and feed on some of the plant matter in the pool. Less than an inch long, the tadpoles were dark, with golden spots. They did not have lungs, as adult toads do; they breathed by taking water into their mouths. The water was sent to the gill chamber, where four pairs of internal gills extracted the oxygen from it. Not for a long while would the tadpoles be able to breathe air.

Slowly but steadily the water continued to evaporate. A week after the storm only three and one-half inches remained. A strong wind came up that afternoon and the water began to evaporate even faster. At this rate, the water would be gone in just a few days—and the tadpoles would die.

Later that evening, when the temperature dropped, a bobcat wandered by. She had been hunting a jackrabbit, but he had escaped. Now she was thirsty. She paused for a moment behind a rock, her pointed ears twitching and shifting direction as they sifted the night sounds. Hearing nothing of interest, she quietly padded over to the rain pool. As she lowered her head and drank, the tiny tadpoles felt the vibrations in their lateral lines—special sense organs for detecting movement—and dove quickly to the bottom of the pool. The bobcat drank for several moments, and when she left the pool was a full inch shallower. Less than two inches remained.

The next day, the wind died down, and very little water evaporated. Perhaps the tadpoles had a chance after all. They were changing rapidly; they already had two hind leg buds, and within a few days the buds would lengthen out, develop toes, and thicken, and the front legs would form.

The tadpoles' tails were shrinking, supplying them with food, for at this stage of their development they could not eat.

Internal changes were taking place, too. The tadpoles' gills were disappearing and their lungs were developing. Now and then the little tadpoles rose to the surface and took a breath of air.

Finally, one morning, they were tiny toads, with only a speck of tail. They hopped out of the rain pool, which now contained less than an inch of water. Together they headed across the sand, leaving the pool behind. Suddenly the sun came out from behind a cloud and the toads hopped into the shade and waited. When the sun disappeared again, they continued on their way. Soon they would find tiny insects to eat; then, like their parents, they would disappear into the sand.

Just fifteen days had passed since the storm.

Two days later the last bit of water disappeared from the rain pool. Only a small damp circle of rocky sand remained. A traveler who happened to walk along that dry stretch of desert would find it hard to believe that just days before, the desert had been alive with toads and tadpoles.

Even if it rained again the toads would remain in the ground. Instinct would not drive them back to the desert for another year—not until the following summer, when the rain clouds would again release their water on the waiting desert.

II

A SUMMER NIGHT IN DEATH VALLEY

The Mohave Desert, which lies in California and Nevada, had not shared in the summer rain. Only three or four inches of rain had fallen over most of this desert during the entire year, almost all of it in winter and spring. Lying in the "rain shadow" of several mountain ranges, the Mohave is one of the driest of deserts. Air carrying moisture toward the Mohave meets the mountain barrier, rises, becomes chilled, and drops its moisture. The far side of the mountains—the desert side—remains dry.

Now it was August. As the morning hours passed, the ground absorbed the sun's radiation and held it, and the heat became more and more intense. The sun shone on the mountains, on the canyons, and on the sand dunes. It shone on rocky ledges in the Funeral Mountains, where small land snails slept in crevices, awaiting the next rain. And it shone on a large stretch of desert sloping downward and westward from the Funeral Mountains, reaching, at its lowest point, two hundred and eighty-two feet below the surface of the sea.

This large stretch of desert was Death Valley, the hottest place in the world.

Here the heat burned into the desert floor and seemed to rise again in ripples. Shortly after one o'clock the air temperature in what little shade existed reached one hundred and twenty-one degrees, and the ground temperature reached one hundred and fifty-eight degrees. This was not at all unusual for a summer day in Death Valley.

A few yards from a sand dune stood a flat, rock-strewn area several hundred yards long. In the ground, beneath a dead-looking shrub, a lizard waited out the heat. A sidewinder, or horned rattlesnake, tightly coiled and quite alert, rested in the sand under a large reddish-brown rock. Neither the sidewinder nor the lizard would venture out just yet. The direct sun could kill them in just a few minutes.

Finally, late in the afternoon, the temperature dropped to a little over one hundred degrees, and the lizard left his burrow and scooted away to search for insects. Not long after sunset the temperature dropped rapidly, as the desert yielded the day's heat. Now the lack of humid air and heavy plant growth worked to the desert's advantage. A little of the heat hit air particles and returned to the desert, but most of the heat escaped, and the temperature plunged to an extent not possible in moist areas.

Now the temperature was in the high eighties. The sidewinder left his shelter beneath the rock and crossed the sand easily, throwing his body forward in a series of loops. He disappeared into the night, leaving a trail of J-shaped parallel lines.

Some minutes later a faint rustling sound came from a sandy mound just under a creosote bush. A pebble rolled down the mound and stopped, and a furry head with a pair of incredibly large, bulging eyes peeked out of an opening. The animal listened; then he scooted out on the sand and paused, sitting up on his huge hind feet and holding his tiny forefeet close to his body, like a kangaroo.

He was a kangaroo rat—but his name did not fit him at all, for he was not a rat, but a squirrel-like rodent. He was, in fact, a very pretty animal; his fur was soft and shiny and well groomed. Including his long, narrow, fur-tufted tail, which was much longer than his body, he was close to a foot in length.

He scurried over to a certain spot near a creosote bush and dug lightly in the sand. With his tiny paws, he picked up some dry seeds and stuffed them into his fur-lined cheek pouches.

The nearest waterhole was many miles away, but this was no problem for the kangaroo rat. Incredibly, he never drank water, even when it was nearby. Dry seeds and dry plants were all he needed to stay alive. Because of a unique type of kidney and other adaptations, he could conserve water and get by on the small amount of water he obtained from his dry food.

Yet, like other desert creatures, he could not stand extreme desert heat. He avoided it by spending the day in a burrow a few feet underground. Even when the ground temperature had neared one hundred and sixty degrees, the kangaroo rat's burrow had remained surprisingly moist and comfortably cool—under eighty-five degrees.

The desert had turned cool, yet there was danger in the night, and the kangaroo rat paused every few moments to listen. He had good reason to be wary: he had come close to death several times. Just the week before, a sidewinder, coiled in ambush in a sandy pit, had lunged at him in the darkness. The kangaroo rat had escaped only by leaping high in the air, out of the sidewinder's reach. Among the most hunted of all desert animals, a kangaroo rat provides not just food but also water to desert carnivores. Although he never drinks, his body, like those of other mammals, is about two-thirds water.

As the night wore on, the kangaroo rat traveled back and forth, gathering seeds and carrying them to the storage chamber in his burrow. Just before dawn, when he was returning home, he caught sight of a small, dark figure at the burrow entrance. Another kangaroo rat had been into his storage room, stealing the seeds he had worked so hard to gather! This was too much!

Furiously, he leaped at the intruder and kicked him hard, sending him flying across the sand. The sneak thief landed with a soft thud and leaped into the air; the instant he hit ground again he kicked sand into his opponent's face.

Now the two animals rushed at each other. Gone was their caution. With their tiny paws, they grabbed each other and spun around and around, squeaking loudly, nipping at each other's fur, and sending sand flying through the air.

For many minutes the battle continued, and the noise carried far in the stillness of the desert night. In the distance, a keen-eared kit fox heard the battle sounds. Her long, sharp-pointed ears twitched; she moved first one, then the other, picking up the exact direction of the noise.

She started rapidly across the desert, her three cubs following.

A few minutes later the fight finally ended. The smaller kangaroo rat, the intruder, broke away, turned, and ran off into the night, and the first kangaroo rat headed toward his burrow. Already dawn was near; a plaintive sound filled the air, for the birds had awakened and were welcoming the morning.

The kangaroo rat was only a few feet from the burrow entrance when he heard a sound that filled him with panic. He turned—just in time to see four pale forms lunge at him! Terrified, he jumped into the air and bounced across the sand on his snowshoe-like feet.

Across the desert he went at tremendous speed. He circled around and around, the kit foxes close behind. For some time, the chase continued; then the kangaroo rat crossed back, leaped head-first into his burrow, and fled down one tunnel, then another, until finally he reached a small chamber, three feet below ground level. Here he huddled, motionless.

But the kit foxes had followed him to his burrow.

The mother kit fox scratched at the sand around the entrance, whimpering softly. The scent of the kangaroo rat was strong; it was a familiar scent, a scent that meant food. The cubs, too, were hungry; they, too, began to dig.

Several minutes later the kit foxes were well into the burrow. But already the kangaroo rat had begun to recover from the chase, and he scooted up through another series of tunnels, away from the side where the kit foxes were digging. Suddenly he burst through the far side of the mound in a blaze of sand—escaping through a hidden emergency exit.

Again, the kit foxes gave chase, and again the kangaroo rat leaped across the sand—jumping high into the air, hitting the ground, then sailing into the air again, covering several feet at a leap. Still the kit foxes followed, darting around rocks and shrubs.

They were only a few feet behind!

High in the air, the kangaroo rat flicked his powerful rudder-like tail and made a ninety-degree turn in mid-air. The unsuspecting kit foxes continued straight ahead as the kangaroo rat bounced away, far to their left, and disappeared headfirst into a shallow escape burrow.

The kit foxes came to an abrupt stop and looked around. The kangaroo rat was no longer ahead of them. They circled frantically, sniffing the sand, trying to pick up the kangaroo rat's trail. Again, and again, they retraced their steps, but they could not pick up the scent.

It was becoming quite light. The kit foxes would have to go to bed hungry, for it was time for them to return to their burrow. The mother kit fox gently nudged her cubs and trotted off; a moment later the cubs followed.

Already the temperature on the desert was well over ninety degrees. The kangaroo rat was beginning to feel the heat. The air in his shallow emergency burrow was cooler than the desert air, but not by much. This burrow would do for a short time, but it was not deep enough to protect him from the heat for long.

Should he try to escape to his deep burrow? Were the kit foxes waiting? If they were, he could no longer escape. He was much too weak.

The temperature continued to rise. Soon the kangaroo rat's body temperature began to rise dangerously. His mouth filled with wetness as his saliva suddenly gushed. It ran out of his mouth, soaking the soft, silky fur on his chest. This was an emergency reaction; it would enable him to stand the heat for a few more minutes—but then, if he did not reach his deep burrow, he would die.

He must try to escape. To stay meant certain death.

Slowly he made his way to the entrance. The desert was quiet. Weak from the heat and from his frantic escape, his little heart beating rapidly, the kangaroo rat crossed the sand to his deep burrow and disappeared.

He was safe.

Again, that sandy stretch in Death Valley was quiet. Both the hunters and the hunted would wait out the day, for they could not withstand the murderous mid-summer heat of Death Valley.

The temperature rose higher and higher as the sun moved across the sky.

III

THE BIGHORN SHEEP

On a cliff in the Funeral Mountains above Death Valley, a bighorn ewe and her four-month-old ram lamb rose from the shade and felt the warmth of the sun hit their backs. The ewe stretched. Slim and graceful, she weighed a little over one hundred pounds and stood less than a yard high at the shoulder. Her coat was tan, her horns small and goat-like.

Pausing a moment, the ewe looked about at the rocky slopes and the shadowy canyon far below. For many weeks this area, several square miles, had been home to her and her lamb. But now they must leave to find a spring, for their water hole had just run dry.

The ewe crossed to a narrow path. This path was familiar to her. It led to a large spring many miles away and several thousand feet below. The summer before, she and her last-year's lamb had watered at the spring with several other bighorn sheep. After several weeks, rain had released them from their need for the spring and they had scattered to the back country to feed on green plants and wildflowers.

The lamb looked about eagerly as he trotted behind his mother. This was his first trip down the path to the spring.

Throwing his head back, he looked high above at the canyon walls. Gaily he whirled, galloped back a few yards, and leaped in the air, landing stiff-legged. Then, with a little snorting noise, he dashed after his mother.

The lamb heard a thunderous “crack” from the canyon below. He rushed to the cliff ledge and sighted down his nose. Far below were two huge bighorn rams, nudging and pushing each other. The rams were half a mile away, but the lamb saw them clearly. Their horns were not like his mother’s. Huge and thick, they curled backward and downward, almost forming a circle.

The rams separated and walked slowly away from each other. But suddenly, when they were twenty feet apart, both whirled. Rising on their hind legs and lowering their heads, they hurled straight at each other. Faster and faster they went. Surely one or the other would break away! But instants later they leaped into the air and hit head on. Again, the tremendous “crack” of their clashing horns echoed through the canyon.

The rams had hit at a combined speed of some sixty miles an hour. Yet, unshaken and unhurt, they landed gracefully on all four feet and stood facing each other, ready for another encounter. Their horns had met perfectly. The swollen bulges of cartilage at the base of the horns had absorbed the shock of the blow.

The rams were not trying to injure each other. Nor were they fighting over a nearby ewe. Going their separate ways from spring to spring in search of ewes, they had met in the canyon and simply followed an urge to charge each other.

In the weeks to come many such encounters would take place, for summer is the height of the rut, as the mating season of sheep is called. Some rams, unable to find another ram to battle, would charge a tree or mountain instead—so great during the rut is the urge to meet another object head-on!

Now, on the ledge high above the canyon, the ewe uttered a low bleat. It was time to move on. The lamb hurried after her.

The sheep moved steadily along, browsing almost constantly. They did little damage to the vegetation, for they nibbled, rather than nipping off entire plants.

Late in the day the ewe and lamb entered a canyon. Ahead, in the lengthening shadows, stood a magnificent ram, his head held high. He spotted the ewe and thundered across the canyon, slowing to a walk as he neared her. Head back and tilted, he edged up to her with short, rapid steps. The ewe turned away, as if to ignore him. But he was not to be ignored! Trotting after her, he gently nipped her side. She tossed her head and walked away.

As the lamb wandered off, browsing, the ewe folded her legs and lay down. Immediately the ram rushed at her.

Leaping to her feet, she ran off. At the crest of a small hill she stopped and turned back toward her lamb. But she could not get to him! The ram headed her off, keeping between her and the lamb and forcing her back.

Trotting off, the ewe looked back over her shoulder. The ram lunged. The ewe broke into a gallop and tried to circle back to the lamb, but it was no use—the ram herded her away.

Finally, with one last backward look, she fled up the steep canyon wall. The ram followed, almost at her heels.

Half an hour later the little lamb got to his feet and looked about for his mother. He saw no one—only the strange canyon, now deep in shadow. Bleating loudly, he broke into a run. He ran in a circle and bleated again, but he heard no deep answering bleat.

He was alone.

He tore up the trail and peered far ahead, but his mother was not there. He pulled his ears back in fright and ran to the opposite side of the canyon. Then, turning around, he backtracked down the trail he and his mother had followed. But his mother was not there either.

It was almost dark when the lamb returned to the canyon where he had last seen his mother. Exhausted, he lay down by the trail, bleated softly, and fell asleep.

He woke in the morning hungry and thirsty and spent the day wandering through the canyon, nibbling on desert holly. Although he had not yet been weaned, he was old enough to live without his mother's milk. Yet he needed water. He followed one trail after another, but he did not find a spring or a water hole. Nor did he find his mother. At night, he bedded down against the canyon wall.

For the first time in his young life, the lamb was in real danger. Life is not easy for a bighorn lamb in Death Valley. Only one or two out of every ten lambs live to be one year old. However, those who live through the first year usually go on to live a full life, dying of old age sometime after ten years.

The dangers of the first year are many. Some lambs fall from cliffs. Some die apparently from an unknown disease marked by a rough coat and a deep, almost constant cough. Some are not strong enough to travel long distances, and die from lack of food or water during the hot, dry season.

And some die because they are lost during the rut-run.

Until now, life had been pleasant for the little lamb. He had been born early in April, somewhat late in the lambing season, which is at its height in January and February. After his birth, the lamb's first awareness was the warm, rough feeling of his mother's tongue gently licking him. He felt her warm body next to his, and heard her strange, soft noises. Moments later he began to nurse.

Soon he tried to stand, but his long, skinny legs buckled and he fell. He slept briefly, nursed, and slept again. When he was only a few hours old he struggled to his feet and stood shakily. But he had little chance to enjoy his new skill, for his mother carefully lowered her head to his—and pushed him over! Gleefully he got to his feet and hobbled toward her. Again, ever so gently, she pushed him over.

As he lay on the ground looking up at her, she leaned over and nuzzled him tenderly. He felt her nearness and the warmth of her breath, and fell peacefully asleep.

Late in the afternoon of the second day he was strong enough to follow his mother a mile up the mountain to bed down for the night. As days passed, the ewe led him through the canyon and up the slopes. No longer did she descend sheer cliffs by "bouncing" from side to side, in the manner of the bighorn. Instead, she took easier routes. And whenever the lamb slipped and fell back while trying to follow his mother up a cliff, she hurried on to a ledge the lamb could easily reach.

One afternoon the lamb caught his foot in a shrub and fell over, bleating loudly. His mother raised her head to look, but she did not help him. He kicked and tugged for several minutes. Finally, he fought his way free, flipped to his feet, and raced up a slope kicking his heels.

Already he was becoming self-sufficient.

The days passed pleasantly. Spring was a time of abundance, a time for play. It was a time for leaping through the wildflowers and playing with other lambs. It was a time for thundering down slopes, jumping on his sleeping mother's side, and galloping away before she was quite aware of what had happened.

When summer came to the mountains, the air was not terribly hot. The ewe and lamb lived at an altitude of over six thousand feet, and the temperature range in the mountains of Death Valley is great. Often the Valley itself sizzled at one hundred and ten degrees, while the temperature high on Telescope Peak, at eleven thousand feet in the Panamint Mountains, was only in the fifties. At the same time the lamb's canyon was pleasantly warm.

Every few days the lamb and his mother visited the only water hole within easy distance—one of several hundred water sources scattered throughout the two million acres of Death Valley National Monument.

But the water hole had dried up, and the ewe and lamb had moved on.

Now, without his mother, the lamb's world had changed. He was alone and far from home. And he was very thirsty. The temperature at this low elevation was one hundred and fifteen degrees. Because of the heat and the lack of green plants, the lamb needed water more often than usual.

In a few days, his thirst was almost unbearable. No longer did he run and leap. He browsed in the early morning and lay in the shade during most of the day, when the sun was at its worst. His sides were sunken, for he was quite dehydrated. He did not know that the spring to which his mother had been leading him was only a few miles ahead. Nor did he know that his mother was free from the ram at last, and for some time now had been searching for him.

An hour or so before dusk the lamb looked across the canyon and caught sight of a familiar form in the distance. Bleating eagerly, he scrambled to his feet and hurried toward it. He had not moved so quickly in days! But a few yards from the bighorn he stopped short. It was not his mother after all—it was a ram. The ram snorted, whirled, and galloped away. The lamb, head down, mouth open, panting very hard now, walked a few steps away and lay down. He did not get up again, but fell asleep where he was.

Late the next day the lamb was nowhere in sight. Stillness had settled over the canyon. Nothing moved. No white rump patch flashed on the rocky slopes.

But a few miles away, several dozen yards from the spring, two gaunt figures paused on the crest of a hill and stood silhouetted against the early evening sky—a young ewe and her ram lamb.

The ewe and lamb crossed slowly to the spring. They were barely able to walk. At each step their skin, wrinkled and shrunken, was pulled tight over their bones. Their ribs and skulls bulged.

Finally, they reached the spring, lowered their heads and drank. After several minutes, their sides bulged with water. The ewe turned away to nibble on grass, but the lamb continued to drink several minutes longer.

Now, as the ewe and lamb stood by the spring, a remarkable change began to take place. The bulges in their sides slowly disappeared. As minutes passed, their coats turned soft and shiny. Their skin and muscles filled out as the cells took in water. Finally, the sheep were no longer thin and bony. They had rehydrated!

It did not seem possible that half an hour earlier both animals had appeared to be near death! But this was no miracle, for bighorn sheep have an incredible ability to recover from dehydration. Without this ability, they probably could not survive in the mountains above Death Valley.

Sometime later the ewe and lamb left the spring to bed down for the night. They climbed easily up the mountain. The ewe did not roam ahead of the lamb. She watched him almost constantly and paused whenever he had trouble keeping up. The lamb did not linger to explore the rocks and ledges on this new trail, but followed his mother closely.

Finally, they reached a high cliff and bedded down very close to each other. By the time the sun disappeared both were asleep.

IV

CLOUDBURST!

Hundreds of miles to the southeast of Death Valley, ominous white clouds began to gather high above Mexico's Sonoran Desert. The clouds had picked up moisture over the Gulf of Mexico. Now hot air rising from the desert hit the cool, moist air in the clouds, and the air began to churn.

A few hours later the sky was purple-black and a thunderhead many thousands of feet thick hung over the desert. From time to time lightning lit the sky. As the day wore on, the layers of ice, snow, and rain in the thunderhead became thicker and heavier. Finally, updrafts could no longer hold the rain within the cloud. Rain began to fall, cold and heavy, on the desert below.

Now the wind turned cold and violent. With a roar, a gust tore across the desert, pushing rain in front of it and yanking plants out of the earth. The wind slammed a huge saguaro. The forty-foot giant toppled slowly and silently; then its seven tons smacked the earth, and the earth shuddered.

The cactus bounced, then lay still. It had grown for well over one hundred years. But it would grow no more.

A few yards from the saguaro several raindrops ran together, forming a trickle. The trickle rolled over the dry, hard ground. At the bottom of a small slope it joined another trickle and added inches to its width. Twisting and rolling, it found its way to a dry wash carved into the ground during thunderstorms long forgotten. Here it met other streams and became a river many yards wide and several feet deep.

The river gained speed and strength and seemed to leap across the desert. It picked up a boulder weighing half a ton and pushed it along as if it were a huge ball of cotton. It reached into burrows. It pulled out rodents and spiders and other creatures and carried them along.

A scorpion rode for an instant on the crest of a wave, then disappeared into the muddy water and did not rise again. A drowned snake hit the branches of a shrub along the edge of the river, dangled a moment, then was swept on.

At a bend in the river a soaked coatimundi, clinging to a dead limb, was flung against the bank; he clawed frantically, struggled up, and dropped, exhausted, on the ground.

The rain continued for several hours before it gradually slowed, then stopped. The river dwindled to a trickle. In the wash lay many dead animals and uprooted trees and plants. In a few short hours, the flood had brought incredible destruction.

Yet as the flood had taken away life it had also brought life. It had picked up seeds of paloverde trees and pounded them against rocks, breaking open the seed coats. It had accomplished what no amount of soaking could accomplish, for only now could the seeds germinate.

The flood had provided water for plants and seeds; now they, in turn, would provide food for animals for many months to come.

Far to the east, new storms began. In the northern Chihuahuan Desert clouds formed over a dry lake—a low area where once, in another age, a small lake stood.

The mud of the dry lake was rock-hard. Caked in it were pebbles and rocks, bits of plants, and many tiny eggs. Now the rain dampened and softened the mud and a pond began to form. When the storm ended almost a foot of muddy water covered the lowest part of the old dry lake.

As the day wore on, very little water seeped into the ground, for the ground beneath the mud was hard. Nor did the water run off, for the lake bed was lower than the surrounding land.

The next morning the water in the shallow pond warmed to over one hundred degrees; and, in the muddy water in the Chihuahuan Desert, life began to stir. The tiny eggs that had been pressed in the dry mud hatched into six-legged larvae with broad heads and elongated bodies. The larvae swam about in the warm water. As hours passed, they grew incredibly fast. As they grew they molted often, and at each molt they lengthened out.

Still very little water sank into the ground. But each day perhaps half an inch evaporated.

One morning a larva molted again and became an adult fairy shrimp—a small crustacean about an inch long, with a large head and a long, segmented body. On his thorax were eleven pair of leg-like appendages. Behind his thorax stretched a flattened, tail-like abdomen. The fairy shrimp was not a true shrimp, but he clearly resembled his relatives of the sea. He swam about rapidly and gracefully in the muddy water.

The fairy shrimp had one small eyespot in the middle of his forehead and two large compound eyes. His compound eyes were dark. They were on stalks that jutted out from the sides of his head.

The sun shone brightly, and the fairy shrimp saw the light from the sun. He swam upside-down with his appendages directed toward the light. He beat his appendages rhythmically and shot through the water, glided gracefully, then zoomed away again.

While the fairy shrimp swam, he breathed through gills at the end of his appendages. His gills took oxygen from the water and released carbon dioxide. As his appendages moved, they set up a current that flowed over his body from front to back.

Some of the current slipped out and passed through bristles on the ends of his legs. The bristles caught tiny bits of plant and animal matter and pushed them into another current along the center of his body. This current moved from back to front, carrying the food into the fairy shrimp's mouth.

Less than a week after the cloudburst, only six inches of water remained in the temporary pond. Perhaps rain would fall again soon—but it did not seem likely. Although a few thunderheads still lingered in the sky, the nearest was over one hundred miles away.

As the fairy shrimp swam, he darted now and then to the mud at the bottom of the pond and rested. Then he beat his appendages and scooted about again, breathing and eating all the while.

Now many others of his kind swam near him. A female darted by, and the fairy shrimp swam to her. He clasped her with his large clasping antennae. The fairy shrimp transferred his spermatozoa to the female. Then she swam away. Other males found other females. Inside the large egg pouch of each female, fertilized eggs began to develop.

Still the water continued to evaporate.

One morning rain clouds appeared on the horizon and moved slowly across the sky. Perhaps the dry lake would again be flooded! These clouds, too, carried moisture from the Gulf of Mexico, but they did not carry much. As they passed over the pond where the fairy shrimp swam, they dropped a little rain. But the rain evaporated high over the desert. Not a drop reached the ground.

One day, less than an inch of water remained in the pond. As the day wore on, this, too, slowly evaporated. The fairy shrimp squirmed and wriggled on top of each other. Their gills could not take oxygen from the air. They had to have water if they were to breathe.

Finally, they no longer had water. One by one, the fairy shrimp died.

The sun shone down on the mud and on the limp bodies of the fairy shrimp and dried them. Caked in the drying mud were thousands of tiny fertilized eggs. The shells of the eggs were thick. As days passed and the mud in the dry lake cracked and curled, the eggs did not dry out. They were well protected. They could remain dormant for many years, perhaps many decades.

Someday another summer cloudburst would flood the dry lake, and in the warm water the dormant eggs would hatch and develop rapidly. In a few days, the dry lake would again be filled with fairy shrimp. Like their parents, the fairy shrimp would swim about gracefully—as long as the desert pond lasted—eating and breathing as they swam, their limbs toward the sun.

V

AUTUMN

Weeks passed, and late summer became early autumn. By October, a few northern desert areas had already experienced frost, but in the western portion of Arizona's Sonoran Desert, known as the Yuman Desert, the first frost was still a month or so away. Days were now pleasant, and nights were quite cool. Soon the snakes and lizards would go into hibernation.

The Colorado River forms the western boundary of the Yuman Desert, and during the hot summer many animals had visited the river every day to drink. Now that it was cool, many came less often.

Still, they came, for along the river's edge were tracks that told of their visits. Pressed in the damp sand were the hoof marks of a herd of pig-like animals called peccaries that had thundered down to the river several days earlier. Nearby were the fresher tracks of mule deer—a doe and her fawn. Scattered along the river bank were the tracks of skunks and coyotes, and in several places the large, sharp claws of a badger had gashed the sand.

Some distance from the river, in the rocky desert to the east, a Gila monster lay in the afternoon sun digesting a small mouse. Although the Gila monster lived near the river, she did not visit the river to drink water. The small mammals and the bird and reptile eggs she was fond of eating provided her with enough moisture.

Once each summer, however, she needed water for another reason. And so, in July, she had gone down to the river and scooped out a four-inch hole in the damp sand. Here she had deposited seven soft-shelled eggs, covered them with sand, and left them. The heat and the moisture had hatched the eggs, and two months ago the young Gila monsters had found their way to the surface and crawled away into the desert.

Near the Gila monster lived many spiders, snakes, and lizards, who neither knew nor cared that the river was nearby. Among them was a coral snake—an extremely venomous but very beautiful snake with bands of red, yellow, and black encircling his body. The coral snake was quite secretive. Like the Gila monster, he was a creature of the night, but unlike the venomous lizard, he would not hunt at all if the nights were too cold.

Not far from the coral snake's burrow was the burrow of a spider. The entrance to the spider's burrow was about an inch and a half wide and lightly covered with silk. At this moment, the spider was just venturing out, and the silk quivered a moment, then tore, as a huge, hairy leg poked through. Another hairy leg followed, then another, until finally eight legs appeared.

The large tarantula's legs spanned almost five inches. They extended a good distance upward and outward from the front part of her body, which was a head and thorax combined. Together the front part of her body and her abdomen were close to two inches long.

The tarantula slowly crawled over to a shrub, stopped under it, and waited, motionless. Several minutes passed. A large black beetle walked by; yet the tarantula did not move. The beetle crossed the clearing several times before he happened to brush against one of the tarantula's legs.

Now, with incredible speed, the tarantula pounced on the beetle, crushed it with her mouth and appendages, and thrust her fangs into the beetle's body, releasing her venom. Soon the softer parts of the beetle became liquid. With her powerful muscles, the tarantula sucked the liquid into her stomach.

For almost all her twelve years of life the tarantula had lived in the burrow under the rock. The area a few feet on each side was her hunting ground. Like other female tarantulas, she rarely left her own hunting area. Ordinarily she remained in her burrow until quite late in the day. Today, however, she had ventured out just a little earlier than usual.

She had recently shed her skin. For many days before molting, she had not eaten, and her appetite, usually excellent during the summer months, was now voracious.

And so, in breaking the usual routine of her hunting, in coming out of her burrow before her habitual hour, she had made a very serious mistake.

From behind a nearby burro bush a huge blue-black wasp saw the tarantula and took to the air. The wasp was close to an inch and a half long, and her very long hind legs trailed behind her body, making her appear even larger. She was the type of wasp known as a tarantula hawk—the largest of the North American wasps.

The wasp landed within a foot of the tarantula. With her orange-red wings flat against her back, she walked toward the tarantula with quick, bold steps. She tapped the tarantula's front legs with her antennae, and the tarantula reared back threateningly, her forelegs in the air. Her fangs snapped into striking position. The tarantula lunged. There was a loud crunching sound, but the tarantula could not seem to sink her fangs into the wasp.

Quickly the wasp slipped her abdomen underneath the tarantula and plunged her stinger into the tarantula's body between the third and fourth right legs. The tarantula struggled desperately, but she could not break away. She fell, rolling over and over on the ground, still in the wasp's clutches.

Suddenly the wasp released the tarantula and walked away, as if the battle were over. But she walked only a few steps before she turned and again headed straight at the tarantula.

The great spider straightened her legs and raised her body high in the air, out of the wasp's reach. The wasp grabbed the tarantula's fourth leg with her jaws. Immediately the tarantula tried to get into striking position, but the wasp pushed her legs against the tarantula's legs and held her off. At the same time, the wasp carefully positioned her abdomen.

Now, lying on her right side, she plunged her stinger in again, this time in a vital spot—the large nerve ganglion on the tarantula's underside, close to the first leg.

Her aim was perfect. The tarantula's legs stiffened, and went limp. She fell to the ground, her motor nerve connections blocked. She would live for some time, but she would never walk again.

Still the wasp continued to inject her venom. More than a minute passed before she pulled her stinger out of the tarantula and walked away.

The wasp's work was not yet finished. She flew over the desert, close to the ground, inspecting her surroundings. She flew over the spot where, beneath the ground, the coral snake slept. She flew beyond a sandy stretch filled with kangaroo rat burrows. Then she landed on a small mound where a low shrub grew.

The sand was fairly soft. With her jaws and legs, the wasp began to loosen the sand just under the shrub. Then she dug, tossing the sand back between her legs. She worked for a long time without stopping. When she finished she had dug a sloping tunnel a foot long. At the very bottom of the tunnel was a large chamber.

The wasp crawled out and examined the entrance. It was marked by sticks and rocks, and by a pile of sand. One by one the wasp carried the sticks and rocks away, and with her feet she raked the sand. When she finished only the opening marked the tunnel.

The wasp circled over the entrance and looked at it from the air. She flew over it again and again, each time making a slightly wider curve. The sandy mound in which she had dug the tunnel was halfway between a creosote bush and a deerhorn cactus. The wasp saw these landmarks, and in her mind, she fixed the location of the tunnel. As in all her actions, she did not think about what she was doing. She was following a pattern of behavior that was beyond her control. But she would be able to find her tunnel again.

Now she flew back to the tarantula. Twice she walked around it; then she took one of the tarantula's legs in her jaws and tugged, pushing her feet into the ground. The spider was very heavy—but not too heavy for the wasp to drag.

For almost half an hour the wasp tugged the spider toward the tunnel a few dozen yards away. It was hard work. She had to detour around each rock and plant. The smallest hill was a huge obstacle, for the tarantula was close to eight times as heavy as the wasp.

At last she reached the tunnel. Leaving the tarantula outside, she crawled into the tunnel, inspected it, and enlarged it in two places. Finally, she dragged the tarantula down the tunnel to the large chamber at the very bottom. She packed dirt tightly around the tarantula's legs, which were pressed close to her body. Even if the tarantula recovered, she would not be able to move.

Now the wasp climbed on top of the tarantula and laid a single egg. She cemented it to a bare spot on the tarantula's abdomen. When she finished she left the tunnel and packed sand over the entrance, closing the grave. Carefully she raked and smoothed the sand until the mound looked exactly as it had when she first saw it.

Already it was almost night, and the tarantula hawk flew away to rest. She would not return to the tunnel again. In the morning, she would feed on nectar, and before long she would hunt again. Again, her quarry would be a tarantula.

No other spider would do.

In three days, the egg she had just laid would hatch into a legless grub and begin to eat the tarantula. The grub would grow rapidly and molt four times. Finally, the tarantula would become an empty shell. Then the grub, two inches long, would spin a cocoon and begin to transform into a pupa.

Had it been a month earlier, the adult wasp would have emerged in a few weeks. But it was now early autumn, and the wasp would spend the fall and winter in the ground. In spring, it would leave its pupal case and cocoon, and, with its strong jaws and feet, it would clear away the sand covering the entrance to the tunnel. Then it would take to the air, a beautiful blue-black creature with bright wings.

And, if the young wasp happened to be a female, she would soon begin her relentless search for tarantulas.

For such is the life of the tarantula hawk.

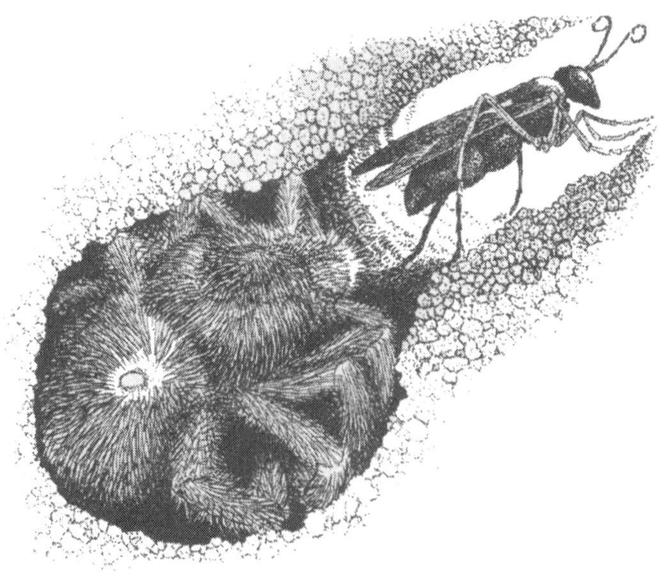

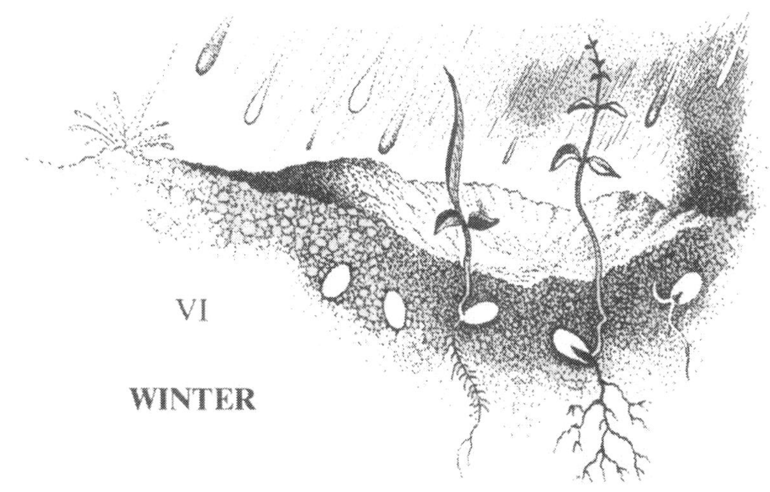

VI

WINTER

Winter settled over the desert in November. It touched the southern deserts gently, turning the days nippy and the nights quite cool. More than once during December the temperature in Death Valley dropped below forty degrees. Twice snow fell on the upper Mohave Desert, glistened briefly on the strange-looking Joshua trees, then melted.

To the north, in the Great Basin or Sagebrush Desert, winter hit hard. Snow fell often and settled on the frozen ground. A foot of snow covered the ground in January, the coldest month of the six-month winter, and several nights the temperature hovered just above zero.

The Great Basin Desert was quiet, for many of the animals, including skunks, bears, and raccoons, had become inactive. Beneath the ground, groups of large desert tortoises slept together in deep tunnels. Snakes hibernated in huge snake dens, some of which contained over a thousand snakes. Yet bobcats still prowled in search of rabbits and squirrels, and mule deer foraged in sheltered valleys.

February was the turning point. The deserts began to warm—the southern deserts rapidly, and the northern deserts gradually.

One day toward the middle of February, masses of black storm clouds gathered over the Pacific Ocean. They moved slowly into California, and climbed high over the mountain barriers. Halfway across Arizona they dropped their moisture, bringing rain to a portion of the Sonoran Desert where the giant cactus grows.

This shower was not at all like the thunderstorm of the summer before. Now the raindrops were small. They fell softly on the paloverdes, on the tiny cacti, and on the huge saguaros. They slid silently down the stems of the creosote bushes and sank into the ground. They fell steadily but gently for almost two days, and then the rain was over.

Yet the story of the rain was just beginning.

The rain had fallen on billions of wildflower seeds just under the surface of the soil. The seeds had not sprouted in December, when over an inch of rain had fallen, for then the ground had been quite cold. Nor had they sprouted during the heavy rains the summer before, for spring wildflowers never sprout after a summer rain, when the ground is hot, nor summer wildflowers after a winter rain, when the ground is cool.

And they had not sprouted three weeks earlier, when less than half an inch of rain had fallen, although the soil surrounding them had become thoroughly wet. In the seed coat of each wildflower seed was a growth inhibitor. Before the seed could sprout, this growth inhibitor had to be dissolved by certain acids formed as water passed downward through the soil.

Now conditions were right. The seed coats became softened and split open; the cells swelled with moisture, and each developing plant pushed a tiny root downward into the soil and began to lengthen.

Many spring wildflower seeds remained dormant. If, by chance, the weather now turned very hot and dry, killing the developing plants before they could mature and drop their seeds, there would still be spring wildflowers in years to come.

Other plants, too, responded to the rain. Shrubs that had dropped their leaves during the dry weeks rapidly took in moisture. Soon they would put out leaf buds. The widespread shallow roots of the saguaros and barrel cacti pulled in water. Soon these great water-storers would become plump and swollen. They would use their water slowly in the dry, warm weeks ahead; if necessary, they could now live many months without rain.

Soon it would be spring on this part of the desert, but the stage was not yet set. If all went well and the wildflowers and other desert plants bloomed, many would need the help of insects if they were to be pollinated. But few insects were now about, for it was still late winter.

Yet, lying dormant under the soil and under rocks and plants were the beginnings of new generations: vast numbers of insect eggs and bee, moth, and butterfly pupae. Even as the rain had moistened the wildflower seeds and stimulated them to germinate, it had also stimulated certain of the insect eggs and pupae. Already they, too, were developing.

It was only February; yet spring was just a few weeks away.

VII

SPRING

Days passed, and the sun warmed the earth. Grass sprang up, the ocotillos became heavy with leaves, and wildflower stems pushed through the soil.

One morning a twig on a small shrub stirred slightly, and a young butterfly appeared. She had just emerged from her chrysalis. Her wings, soft and wet, were folded on her back.

With her wings toward the ground, the butterfly gripped the twig firmly. She had many things to do, and she must do them quickly. She began to swallow large amounts of air, and to take air into her body through openings called spiracles. As moments passed, pressure built up inside her body. She began to expand.

Her muscles contracted, forcing blood into her wing pads. Her wings grew larger. The membranes of their upper and lower surfaces came closer and closer together until the wings could grow no more.

The butterfly moved her wings gently up and down. They

were still limp and moist. She fitted and locked together the two parts of her slender feeding tube, called a proboscis, coiling and uncoiling it as she did so.

Slowly her wings and her outer skeleton hardened.

She fluttered her wings again. Almost an hour had passed since she had emerged from her chrysalis. Now her wings were stiff and strong. She was ready!

Over and over she raised and lowered them—slowly at first, then faster. Faster! She let go of the twig, and her wings carried her into the air. She was flying!

She soared into the air, higher and higher. She felt the warmth of the sun. The earth beneath her was bright with color, and the air was sweet with the smell of flowers.

The desert was in bloom.

The butterfly came to rest on a creosote bush, and her eyes took in the world around her. The desert was almost covered with flowers—yellow flowers, white flowers, and clusters of red, blue, orange, and delicate pink.

The butterfly saw many of the colors, but she did not see them as we see them. The flowers we see as blue were different shades to her. The flowers we see as red were either ultraviolet or black.

She saw nearby flowers somewhat clearly, but distant ones were fuzzy.

Suddenly she saw the quick movement of a cactus wren as it flew to a nearby cholla. This she saw very well, for her compound eyes were well suited for detecting motion.

Then she caught sight of a cluster of orange flowers several yards away. This was more important than anything she had yet seen. She flew to the orange flowers. When she landed, taste organs on the soles of her feet contacted sweet nectar. Immediately her proboscis uncurled and reached deep inside the flower, and she pumped up nectar.

It was her first meal as a butterfly. For the rest of her life she would eat nothing but nectar.

Pollen clung to the butterfly's legs. When she flew to another orange flower the pollen dropped inside the flower.

For many days, she and others of her kind would transfer pollen from flower to flower. Certain solitary bees, attracted only to certain blue flowers, would pollinate these. In the days to come, other types of bees would emerge from their underground cells or from their cells inside trees at precisely the right time to pollinate other flowers just as they bloomed.

Such is the timing of spring on the desert.

From a saguaro branch, high above the orange flowers, a Costa's hummingbird watched the butterfly. The hummingbird was only a little over three inches long, not much larger than the butterfly. His back was a lovely greenish-bronze, his throat and forehead bright purple.

The hummingbird had been in the desert only a short time, for he had spent the winter in Mexico. Now, from his perch in the saguaro, he looked down at the desert. He saw the orange flowers beneath him. On his right, he saw a pair of ground squirrels chasing each other around a yucca. Then he raised his head and looked far into the distance. His eye muscles changed the shape of his lenses and he focused on a far slope.

There he saw bright red ocotillo blossoms waving gently in the air. He saw them clearly, for his eyesight was remarkably keen. Unlike the butterfly, he saw the flowers as red. He was strongly attracted to this color.

He winged his way rapidly toward the red flowers. He flew over the saguaros, over a mesquite thicket, and over a small hill covered with white flowers. He reached the ocotillo blossoms and hovered before one of them. His wings beat incredibly fast—several dozen times a second —and he became a blur. A soft humming sound filled the air.

With his narrow tube-like tongue, the hummingbird reached deep inside the blossom and sipped the nectar. Tiny bits of pollen clung to his head, where he brushed against the inside of the flower.

For several hours, he continued darting from flower to flower, sipping nectar. As he did, he picked up pollen grains from many flowers and dropped them into many others.

Early in the afternoon a monarch butterfly passed by the ocotillo slope. She flew gracefully and easily, raising and lowering her black and orange wings several times, then gliding. She showed no interest in the ocotillos as she passed. Suddenly, however, she veered to the left. She had detected the odor of milkweed.

She could not see the milkweed plants, yet she knew they were nearby. She followed the odor until finally, beyond a small valley of purple flowers, she found the plants.

She dropped onto a milkweed leaf, crawled to the underside, and paused a moment, spreading her wings slowly. Then, turning so her abdomen was close to the center of the leaf, she laid a single pale green egg. With a gland near the end of her abdomen she secreted a cement and fastened the egg securely to the leaf.

In a few days, a small greenish caterpillar would hatch from the egg. He would remain a caterpillar for almost two weeks, eating nothing but milkweed leaves during this time.

The monarch flew on. She seemed to drift aimlessly from right to left, yet she was heading steadily northward. Already she had traveled far from her winter home in Mexico.

As she flew, she would pause often to place a single egg on a milkweed leaf, for her migratory flight followed the spring growth of milkweed plants. She would die before she reached the limits of the milkweed growth in Canada, but many of her offspring would reach Canada before summer's end. In the fall, some would join huge flocks to make the return flight to Mexico.

A Gila woodpecker saw the monarch fly by, but he paid no attention. The monarch's pattern and coloring were quite conspicuous, and the woodpecker, like other birds, would not try to eat a monarch. The milkweed diet of the caterpillars gave them and the adult monarchs a very bitter taste.

The woodpecker was quite busy. He had found dozens of very tiny green caterpillars crawling on the saguaro. Unlike monarch caterpillars, these did not taste bitter, and the woodpecker was eating them as fast as he could. He was having no trouble this year finding insects. The year before, rainfall had been light, few wildflowers had germinated, and many insects had remained dormant. Even though some of the thrashers and cactus wrens had not nested at all and others had produced fewer eggs than usual, many young birds had still starved to death.

This, however, would be a good year. Already many birds were pairing off, and a pair of cactus wrens had almost finished building a nest among the sharp spines of a nearby cholla.

And this would be a good year for other desert creatures. The lizards, out of hibernation weeks earlier, were scooting about, eating insects. The desert rodents, soon to bear young, were nibbling constantly and growing plump—providing food for kit foxes, bobcats, and snakes.

Spring, however, is all too short on the Sonoran Desert. Already each afternoon the air was becoming quite warm. Soon the saguaros, barrel cacti, and other desert plants would bloom, but the time of the spring wildflowers was almost over. Within a few weeks they would drop their seeds and die. Many insects, too, would die.

Yet, like the wildflowers, the bees and the butterflies would leave behind them the beginnings of a new generation. And again, in a year, or perhaps two or three, ample winter rain would fall on the wildflower seeds, on the plants and shrubs, and on the dormant insects.

And spring would come again to the land of the saguaro.

VIII

THE SPRINGTIME OF THE BIRDS

Across the deserts, the birds had felt the change from winter to spring. Stimulated by hormones, their testes and ovaries had grown larger. And the springtime of the birds began.

Early one cool, sunny morning dozens of sage grouse gathered on a barren flat in the Great Basin Desert. One particularly splendid bird, well over two feet in length, spread his huge tail of twenty pointed feathers and waddled through the group, straight and proud.

He stopped near a group of hens. Breathing in and out several times, he filled his lungs and the air sacs on the side of his neck. The white feathers surrounding the air sacs snapped up, almost hiding his head, until he looked like a plump matron with a huge white collar.

Suddenly he raised his wings, tightening the skin between the air sacs. Air rushed out with a loud "plop!" The sound carried far across the breeding ground. As the white feathers settled down again, the bare spots over the air sacs disappeared. The sage grouse strutted stiffly across the flat, flapping his wings, his sharply pointed tail feathers erect and widely spread.

He stopped near a hen and peered at her, as if to see what she thought of his remarkable display. Her response was important to him. Unlike many birds, he would attract as many females as possible. He had already fought other males and would continue to do so, to protect his right to collect a harem. Some of the less aggressive males would have only one mate; others would have none.

Courtship display and mating would continue for many weeks. Eventually each female would scratch out a nest near a stream or spring, perhaps under a sage bush. She would lay six to nine eggs. Alone, she would care for the eggs, and later for the young.

But that was yet to come, for this was early spring. And throughout the barren flat in the land of sagebrush, greasewood, and saltbush, the hens watched as the male sage grouse strutted, danced, spread their tails, and inflated and deflated their air sacs. “Plop! Plop! Plop! Plop!”

On this splendid morning, spring came to the sage grouse on the Great Basin Desert.

Nowhere in the desert did the springtime of the birds arrive with a greater flurry than in southeastern Arizona. Here, more land birds nest, mile for mile, than in any other area in the country. From the sweet notes of the Palmer’s thrashers to the harsh chirps of the cactus wrens, the songs and chatters of countless birds filled the saguaro forests.

With four rapid high-pitched yips, a Gila woodpecker swooped to a saguaro. Her black and white stripes flashed briefly, then she disappeared into a hole in the saguaro stem. The Gila woodpecker had recently pecked out the hole. The saguaro sap had hardened around it, protecting the tons of water inside the saguaro from evaporation. The sap had formed a firm, dry lining. Now the chamber was an excellent birdhouse.

Many nearby saguaros contained several abandoned woodpecker holes. These ready-made homes had been chosen by other birds as nesting sites, and now the forest was alive with flycatchers, sparrow hawks, bluebirds, English sparrows, and purple martins, all popping in and out of old Gila woodpecker or gilded flicker holes. Inside other holes, screech owls and sparrow-sized elf owls slept, awaiting the night.

Other creatures, too, found the old woodpecker homes handy. One hole in the tall saguaro, far below the Gila woodpecker's current nest, had previously housed a succession of tenants—several lizards, a snake, various rats and mice, and a scorpion.

A cactus wren flew by the saguaro and disappeared into a cholla cactus. Here, beneath the abandoned cup-shaped nest of a house finch, the cactus wren had hung her woven, flask-shaped nest. Her young were protected from most predators by the sharp cholla spines.

A red-tailed hawk circling high above the saguaro caught sight of a quick movement on the ground. Plunging, he sank his talons into a ground squirrel. He shot into the air again and, clutching his prey firmly, headed toward his young in a distant nest—a pile of creosote branches high in the arm of a saguaro.

Many minutes after the hawk had disappeared, a group of sage thrashers landed near the saguaro and ran rapidly across the ground, their tails in the air. The thrashers were only passing through the saguaro forest. They were on their way to sagebrush country in the Great Basin Desert.

As the day wore on, other birds flew about in the saguaro forest—tiny verdins, sparrows, rock wrens, mockingbirds, doves, shrikes, goldfinches, and Say's phoebes. Some carried grass, seed pods, or other nesting material in their beaks; others carried insects to feed their young.

Not until the sun had disappeared did the birds rest.

As spring broke through on the Mohave desert, a Scott's oriole perched high in a Joshua tree and sang his sweet, meadowlark-like song. Nearby was his mate, busy weaving a basket of grass and yucca to hang on the thorny Joshua leaves.

Hawks, thrashers, and shrikes gathered nesting material and flew to and from the Joshua trees, for these tree lilies were the only tall trees for miles. The Joshua trees are as vital to the tree-nesting birds of the Mohave as the saguaros are to the tree-nesting birds of the Sonoran Desert.

Not far away, a roadrunner ran rapidly down a dry wash. Without slowing down, the huge, dark bird leaped a foot into the air to grab a bee. She swallowed the bee, made a sharp turn, and disappeared around a mesquite bush.

Several yards down the wash a lizard moved out into the sun. The lizard was sluggish, for the air was cool. He basked in the sun, taking in its radiation, until finally his body temperature was much higher than the air temperature. Now he could move easily. He scooted into the wash.

There went a fly! Off went the lizard. Suddenly something dark streaked from behind a bush and pounced on the lizard.

The sharp eyes of the roadrunner miss very little!

With the lizard in her beak, the roadrunner ran down the wash. She rarely flew, even when she was in a hurry. Darting around a bush, she plunged into a mesquite shrub.

The roadrunner had no need for Joshua trees or saguaros. Her nest was in the mesquite shrub, a few feet above the ground. In the nest were two blue-green eggs, two young featherless birds, and an older bird, almost ready to leave the nest. As the birds squawked and opened their mouths wide, the roadrunner jammed the entire lizard, head first, down the throat of one of her young. Then she ran off to hunt again.

It was fortunate that the roadrunner and her mate did not have a whole nest full of young to feed. Just keeping themselves fed was a difficult task, for roadrunners are almost constantly hungry. In addition to the bee, the roadrunner had recently gulped down several grasshoppers, two wasps, a mouse, and a scorpion. She had also attacked a rattlesnake and had fought with it, dodging its strikes, until she was able to jab her beak into the snake's brain.

But now she was hungry again—and so were her young.

The days of spring whirled by. Throughout the land of the Joshua tree and the land of the saguaro, throughout the desert flats and slopes, valleys and washes, nestlings squawked and flapped their wings and opened their mouths wide for food.

Across the deserts, roadrunners ran, hawks circled, and

Gila woodpeckers snapped up insects and popped in and out of saguaro holes.

Gradually the young birds became strong and fully feathered and left their nests to follow their parents about on the desert. As the young birds watched, they learned. The young hawks learned to circle, spot their prey, and plunge. The Gila woodpeckers learned to find cactus fruits and berries, and to snap insects in mid-air. The seed-eating birds learned where to fly to find the water they needed to live on the desert.

The young birds were on their own. The springtime of the birds was over.

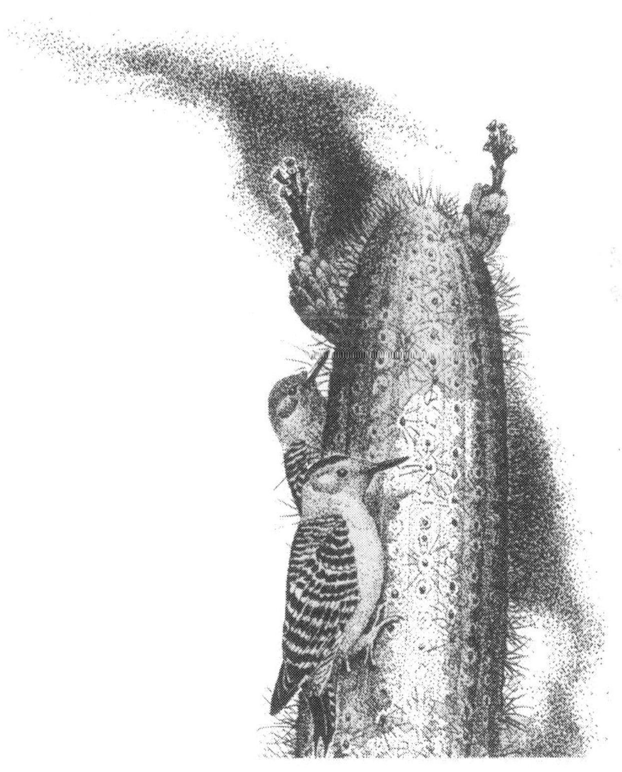

EPILOGUE

July came to the Sonoran Desert. Late one afternoon the sky darkened, and a cool wind whipped across the slopes and flats. As lightning lit the sky, a lizard scurried into his burrow. A Gila woodpecker shrieked, swooped to a saguaro, and disappeared.

Finally, the plop of raindrops mingled with the sounds of thunder and wind.

Deep in the ground a Couch's spadefoot toad sensed the rain and stirred. For a year he had huddled in his burrow, through the heat of summer and the cold of winter, through the rains of December and February, and through the springtime of the birds.

But now he made his way toward the surface of the desert. This was the summer rain. A year was over; another was about to begin.

BIBLIOGRAPHY

Edmondson, W. T., Editor, *Freshwater Biology*. John Wiley and Sons, 2nd. ed., 1959.

Howes, Paul G., *The Giant Cactus Forest and Its World*. Duell, Sloan and Pearce; Little, Brown, 1954.

Jaeger, Edmund C., *The California Deserts*. Stanford University Press, 4th. ed., 1965.

Jaeger, Edmund C., *The North American Deserts*. Stanford University Press, 1957.

Krutch, Joseph Wood, *The Desert Year*. William Sloane Associates, 1952.

Krutch, Joseph Wood, *The Voice of the Desert, A Naturalist's Interpretation*. William Sloane Associates, 1955.

Leopold, A. Starker, *The Desert (Life Nature Library)*. Time Incorporated, 1961.

Oliver, James A., *The Natural History of North American Amphibians and Reptiles*. D. Van Nostrand Company, Inc., 1955.

Pearson, T. Gilbert, Editor, *Birds of America*. Garden City Publishing Company, Inc., 1936.

Peterson, Roger Tory, *Birds Over America*. Dodd, Mead and Company, rev. ed., 1964.

Schmidt-Nielson, Knut, *Desert Animals: Physiological Problems of Heat and Water*. Oxford University Press, 1964.

Smyth, H. Rucker, *Amphibians and Their Ways*. Macmillan Company, 1962.

Stebbins, Robert C., *Amphibians and Reptiles of Western North America*. McGraw-Hill Book Company, 1954.

Sutton, Ann and Myron, *Life of the Desert*. McGraw-Hill Book Company, 1966.

Welles, Ralph E. and Florence B., *The Bighorn of Death Valley*. United States Government Printing Office, 1961.

THE AUTHOR AND THE ARTIST

The author and the illustrator have several characteristics in common. Both grew up in Illinois, and graduated from Northwestern University. Both changed career paths somewhat, in midlife, moving into a different subject area (see *Transitions,* below).

But, most important: the deep love of nature that the author and the illustrator share is so evident in their work. The art of the word and the visual art in *A Year on the Desert* complement each other, bringing nature to the reader, and giving special meaning to the stories in this book.

About the Author

Barbara Goodheart is an award-winning author whose interest in nature and wildlife dates back many years. As a child living on Chicago's south side, she often took long bike rides to visit the city's zoos and museums. But her favorite times took place at the family's summer cottage in Indiana.

At the cottage, Barbara loved to explore the weedy areas near the lake. She'd catch frogs, then let them go; and she'd watch snakes and spiders, always from a respectful distance. She would dig for earthworms, then take the family's clunky wooden rowboat out on the lake, in hopes of catching bluegills, bass, or sunfish for dinner. A free spirit, she went barefoot from morning to night, ignoring her mother's warnings—until the day she stepped on a bee and was stung.

Barbara's interests in the outdoors continued while she was growing up, and long after the family sold the cottage in Indiana. She graduated from Northwestern University with a degree in—no surprise—Biology.

Barbara's husband, Clyde, who was her lab instructor in Comparative Anatomy at Northwestern, shares her interest in biology and the outdoors. While their son and two daughters were growing up, the family often tent-camped in rather primitive surroundings. Their son still recalls the night a skunk sauntered in an open end of his tiny "pup" tent. He knew enough to remain calm and avoid moving, so, much later, the skunk meandered out the other end of the tent, without incident. Barbara watched from the far end of the campground, the only person terrified during the skunk's visit.

Transitions. During her early career as an author, many of Barbara's magazine articles had a natural history or wildlife theme. But, like Mel, her artist-collaborator, she eventually changed her career path. She became more heavily involved in medical writing.

Her first medical book, *Diabetes*, earned reviewers' recommendations in *Booklist, Book Review Forum,* and *The Book Report,* and won Honorable Mention in an American Medical Writers Association competition.

Currently Barbara is a contributing writer at Addiction Treatment Forum *(atforum.com)*. She is also collaborating with her husband, Clyde, an MD, on a series of diabetes books for patients.

But *A Year on the Desert* has never been far from her mind. During a vacation trip, she spotted a copy of her book in a gift shop in a national park. And one day some time ago her son called to say that her grandson had come home from school with the news that "our class was reading a book about the desert, and guess what, the lady who wrote it has the same name as my grandmom!"

Her son didn't miss a beat. "Guess what, the lady who wrote it *is* your grandmom!"

About the Artist

Milford "Mel" Joseph Hunter III overcame the challenges of a difficult childhood to graduate from Northwestern University and go on to become a famous artist. He illustrated the works of well-known science fiction authors, among them Isaac Asimov and Robert A. Heinlein. He also provided scientific illustrations for the Pentagon, the National Aeronautics and Space Administration (NASA), the Massachusetts Audubon Society, and Hayden Planetarium.

It was quite a few years after his graduation from Northwestern, while he was working as a draftsman, that Mel decided to become an artist. He moved to New York City and began teaching himself the techniques of art, in his spare time. A few years later he sold his first color cover, to *Galaxy Science Fiction* magazine.

In 1960, Doubleday published *The Missilemen,* a book with black-and-white photos Mel had taken during visits to U.S. rocket and missile sites. The book depicted the world of rocket scientists and engineers of the early space age. In 1961, another of Mel's books, *Strategic Air Command,* was awarded the Aviation Writers' Association highest honors.

Mel was nominated for the Hugo Award for Best Professional Artist for the years 1960 through 1962. The prize is awarded for works related to science fiction or fantasy.

Transitions. Eventually Mel's interests took a turn—from space technology to natural history and our vanishing wildlife.

According to the Artworks and Auctions website, Mel "moved to Vermont to paint the land and its wildlife, creating magnificent works of art in oil paintings and original art graphics, and to write and illustrate books for children on what nature and the Earth are really all about."

Mel then created one hundred and thirty watercolors of Birds of the Northeast for the Abercrombie & Fitch Galleries and the Massachusetts Audubon Society. He also contracted with World Publishing Co. for a series of thirteen ecology books for children.

In 2004, the year of Mel Hunter's death, his widow, Susan Smith-Hunter, opened the Smith-Hunter Studio and Gallery, where she displayed and offered for sale Mel's lithography and her sculptures.

Mel's interests in rockets and space reemerged before his death in 2004. His final wish was to have his ashes launched into space. The New Frontier Flight, a private launch coordinated by Space Services Inc., obliged, carrying out Mel's last wish on May 22, 2012.

Made in the USA
Columbia, SC
23 March 2020

89837170R00045